My Favorite COLOR

Orange

Kaite Goldsworthy

AV2

www.av2books.com

My Favorite COLOR

Orange

Contents

What Is This Color?

Have I seen it before?

It is orange
I see!

Will you help
me find more?

I see an **orange** chair.

I see an **orange** wall.

Is there **orange** in your home?

Is it big or small?

Orange is the color of some foods people eat.

Which **orange** foods would you like for a treat?

Orange toys can be so much fun!

Which **orange** toy is your favorite one?

I see **orange** flowers.

I see **orange** trees.

Do you see **orange** leaves as they move in the breeze?

I see an **orange** tiger.

I see a bird, too.

Finding **orange** animals is not hard to do.

Playgrounds are fun places to play.

I see **orange** things I could climb on all day.

Can you see **orange** at school?

Will you help me look?

I see **orange** pencils.

I see an **orange** book.

Orange can mean warm.

It can mean safety, too.

Orange is for Halloween.

What does orange mean to you?

Find where these **orange** things belong in this book.

Go back through the pages and have a close look!

KEY WORDS

Research has shown that as much as 65 percent of all written material published in English is made up of 300 words. These 300 words cannot be taught using pictures or learned by sounding them out. They must be recognized by sight. This book contains 63 common sight words to help young readers improve their reading fluency and comprehension. This book also teaches young readers several important content words, such as proper nouns. These words are paired with pictures to aid in learning and improve understanding.

Page	Sight Words First Appearance
4	before, have, I, is, it, this, what
5	find, help, me, more, see, will, you
6	an
7	big, home, in, or, small, there, your
8	eat, foods, of, some, people, the
9	a, for, like, which, would
10	be, can, much, so
11	one
12	trees
13	as, do, leaves, move, they
14	too
15	animals, hard, not, to
16	are, places, play
17	all, could, day, on, things
18	at, look, school
19	book
20	mean
21	does

Page	Content Words First Appearance
4	color
5	orange
6	chair, wall
9	treat
10	toys
12	flowers
13	breeze
14	bird, tiger
16	playgrounds
19	pencils
20	safety
21	Halloween

Published by AV2
276 5th Avenue, Suite 704 #917
New York, NY 10001
Website: www.av2books.com

Library of Congress Cataloging-in-Publication Data

Names: Goldsworthy, Kaite, author.
Title: Orange / Kaite Goldsworthy.
Description: New York, NY : AV2, [2022] | Series: My favorite color | Audience: Grades K-1
Identifiers: LCCN 2021005141 (print) | LCCN 2021005142 (ebook) | ISBN 9781791134228 (library binding) | ISBN 9781791134235 (paperback) | ISBN 9781791134242 (ebook other)
Subjects: LCSH: Orange (Color)--Juvenile literature. | Colors--Juvenile literature.
Classification: LCC QC495.5 .G65 2022 (print) | LCC QC495.5 (ebook) | DDC 535.6--dc23
LC record available at https://lccn.loc.gov/2021005141
LC ebook record available at https://lccn.loc.gov/2021005142

Printed in Guangzhou, China
1 2 3 4 5 6 7 8 9 0 25 24 23 22 21

022021
100920

Art Director: Terry Paulhus **Project Coordinator:** Sara Cucini

AV2 acknowledges Getty Images as the primary image supplier for this title.